A Secondary Mathematics Core Curriculum

SUPPLEMENTS

Developed by high school teachers who used them in their classrooms, these Supplements have been prepared to complement **MATH** *Connections*: A Secondary Mathematics Core Curriculum, Year 1. These problem solving and mathematics skill activities are designed to give students the opportunity to enhance their experiences with **MATH** *Connections*.

It is suggested that **MATH** *Connections* teachers assign these Supplements according to the individual needs of their students. With an average of 15 Supplements per chapter of Year 1, these Supplements may be used to increase student understanding of specific concepts and skills, to provide support materials for students who have been absent from class or to accommodate students who enter the **MATH** *Connections* program midstream.

Robert Fallon
Bristol Eastern High School

CONTRIBUTORS

Thomas Alena
Talcott Mountain Science Center, Avon

Kathleen Bavelas
Manchester Community-Technical College

Carl Bujaucius
Manchester High School

William Casey
Bulkeley High School, Hartford

Frank Corbo
Staples High School, Westport

Helen Crowley
Southington High School

Sharon Heyman
Bulkeley High School, Hartford

Helen Knudson
Choate Rosemary Hall, Wallingford

Mary Jo Lane
Granby Memorial High School

Lori White Moroso
Beth Chana Academy for Girls, Orange

Joanna Shrader Panning
Middletown High School

John Pellino
Talcott Mountain Science Center, Avon

Linda Raffles
Glastonbury High School

Stephen Steele
Hall High School, West Hartford

Pedro Vasquez, Jr.
Multicultural Magnet School, Bridgeport

Thomas Willmitch
Talcott Mountain Science Center, Avon

Legend

Each Supplement begins with the Teacher Commentary (upper left corner), followed by the student pages. The first page of each is labeled with the Supplement Number in the form of 4 - 1; 4 refers to the Chapter in the Student Edition of the **MATH** *Connections* text, 1 refers to the first Supplement for Chapter 4. On the first teacher page under the title, the Supplement is referenced to the specific section in the student text of **MATH** *Connections*.

In the "To the Teacher" paragraph, the objective of the Supplement is stated along with mention of any additional information or materials needed for that Supplement. On the teacher pages only, the Student Supplement is in regular type and the answers appear in bold type. "**MATH** *Connections* 1" (bottom left corner) refers to Books 1a and 1b of the student text of **MATH** *Connections*. Each page of each Supplement is numbered (center bottom of the page). For example, T-1 is page 1 of the Teacher Commentary; S-1 is page 1 of the Student Supplement.

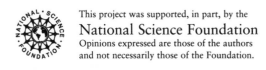

This project was supported, in part, by the
National Science Foundation
Opinions expressed are those of the authors
and not necessarily those of the Foundation.

ISBN 1-891629-93-X (Year 1B Hardcopy)
ISBN 1-891629-94-8 (Year 1B Hardcopy/Electronic)
ISBN 1-58591-003-1 (Complete Set Hardcopy)
ISBN 1-58591-004-X (Complete Set Hardcopy/Electronic)

Supplement No. 5-1

The Educated Guess
(Suitable for use with Section 5.1 of Year 1)

To the Teacher: *This exercise gives students practice determining the number of guesses needed in a given situation to reach the correct answer. Powers of two play a big role.*

1. What would your first guess be if you were playing the guessing game and you know the unknown number was between the following two numbers?

    (a) $1 \leq X \leq 12$
        $X \leq 6$

    (b) $3 \leq X \leq 25$
        $X \leq 14$

    (c) $11 \leq X \leq 61$
        $X \leq 36$

    (d) $20 \leq X \leq 100$
        $X \leq 60$

    (e) $163 \leq X \leq 212$
        $X \leq 187$ or $X \leq 188$

    (f) $475 \leq X \leq 900$
        $X \leq 687$ or $X \leq 688$

2. If an unknown number is between each of the following two numbers, then how many guesses will be needed to assure yourself of finding the missing numbers?

    (a) $1 \leq X \leq 6$
        3 guesses

    (b) $1 \leq X \leq 50$
        6 guesses

    (c) $1 \leq X \leq 295$
        9 guesses

    (d) $1 \leq X \leq 1000$
        10 guesses

(e) $35 \leq X \leq 135$
7 guesses

(f) $268 \leq X \leq 1461$
11 guesses

(g) $1158 \leq X \leq 1161$
2 guesses

(h) $3691 \leq X \leq 5722$
11 guesses

3. It took Willie six guesses to find a missing number while playing the guessing game. The largest number in the range of numbers was 143. What is the smallest number?
79

4. It took June eight guesses to find a missing number while playing the guessing game. The smallest number in her range of numbers was eighteen. What should June's first guess be?
X ≤ 146

5. It took Wally five guesses to find a missing number while playing the guessing game. His first guess was $X \leq 24$. What is the largest number in the range of numbers?
40

6. It took Tiesha eleven guesses to find a missing number while playing the guessing game. Give a possible range of numbers that Tiesha was selecting from.
Answers will vary. One possible solution is 1 ≤ X ≤ 2048

7. Juan ran thirty-two laps around a track on day one. On day two he ran one-half the amount of laps he ran on day one. On day three he ran one-half the amount of laps he ran on day two. He continued this pattern until he ran one-quarter of a lap. How many days did it take Juan to do this?
8 days

Supplement No. 5-1

The Educated Guess

1. What would your first guess be if you were playing the guessing game and you know the unknown number was between the following two numbers?

    (a) $1 \leq X \leq 12$

    (b) $3 \leq X \leq 25$

    (c) $11 \leq X \leq 61$

    (d) $20 \leq X \leq 100$

    (e) $163 \leq X \leq 212$

    (f) $475 \leq X \leq 900$

2. If an unknown number is between each of the following two numbers, then how many guesses will be needed to assure yourself of finding the missing numbers?

    (a) $1 \leq X \leq 6$

    (b) $1 \leq X \leq 50$

    (c) $1 \leq X \leq 295$

    (d) $1 \leq X \leq 1000$

    (e) $35 \leq X \leq 135$

    (f) $268 \leq X \leq 1461$

    (g) $1158 \leq X \leq 1161$

    (h) $3691 \leq X \leq 5722$

3. It took Willie six guesses to find a missing number while playing the guessing game. The largest number in the range of numbers was 143. What is the smallest number?

4. It took June eight guesses to find a missing number while playing the guessing game. The smallest number in her range of numbers was eighteen. What should June's first guess be?

5. It took Wally five guesses to find a missing number while playing the guessing game. His first guess was $X \leq 24$. What is the largest number in the range of numbers?

6. It took Tiesha eleven guesses to find a missing number while playing the guessing game. Give a possible range of numbers that Tiesha was selecting from.

7. Juan ran thirty-two laps around a track on day one. On day two he ran one-half the amount of laps he ran on day one. On day three he ran one-half the amount of laps he ran on day two. He continued this pattern until he ran one-quarter of a lap. How many days did it take Juan to do this?

Supplement No. 5-2

Visualizing Statistics
(Suitable for use with Section 5.1 of Year 1)

To the Teacher: *This Supplement provides experience predicting, graphing and analyzing real world situations.*

Homerun Race

Games Played	Player A	Player B
20	11	7
40	19	14
60	56	23
80	31	29
100	39	38
120	47	48
140	56	55
160	?	?

1. Two major league baseball players were competing for the homerun title. The chart above gives their totals after a given number of games. Answer the following questions.

   (a) Graph the data for player A. Are the points close to forming a straight line?
   Yes

   (b) Graph the data for player B. Are the points close to forming a straight line?
   Yes

   (c) Do the graphs of the two players intersect somewhere? If so, graph the intersection point.
   Yes, (131.8, 51.5)

   (d) After how many games did they have the same number of homeruns? How many homeruns?
   Approximately 132 games, 52 homeruns.

   (e) How many homeruns will each player have after 160 games?
   Player A – 160 games, approximately 62 homeruns
   Player B – 160 games, approximately 63 homeruns

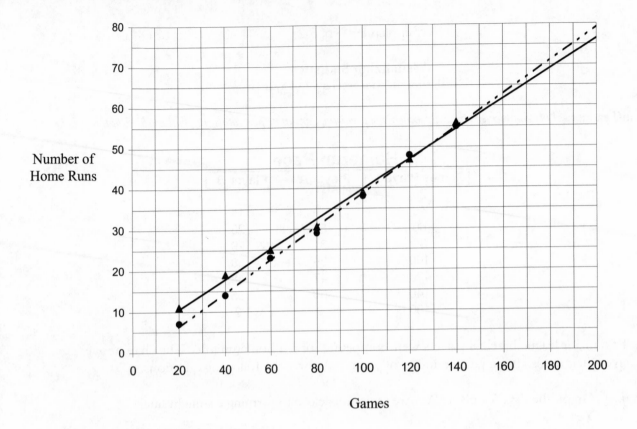

Name _____   Date _____

Visualizing Statistics

Homerun Race

Games Played	Player A	Player B
20	11	7
40	19	14
60	56	23
80	31	29
100	39	38
120	47	48
140	56	55
160	?	?

1.   Two major league baseball players were competing for the homerun title. The chart above gives their totals after a given amount of games. Answer the following question:

   (a) Graph the data for player A. Are the points close to forming a straight line?

   (b) Graph the data for player B. Are the points close to forming a straight line?

   (c) Do the graphs of the two players intersect somewhere? If so, graph the intersection point.

   (d) After how many games did they have the same number of homeruns? How many homeruns?

   (e) How many homeruns will each player have after 160 games?

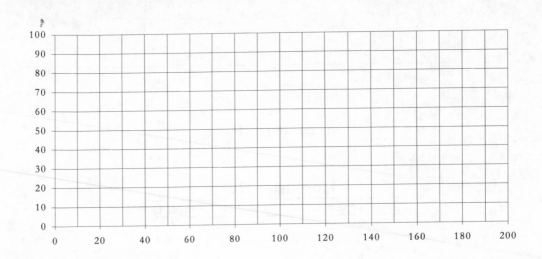

Supplement No. 5-3

The Guessing Game
(Suitable for use with Section 5.2 of Year 1)

To the Teacher: *This Supplement provides students with the opportunity to use the guessing algorithm to determine the roots of numbers.*

.   Find the following roots to 4 places by using the guessing game.

(a) $\sqrt{7} = \mathbf{2.6458}$

(b) $\sqrt{11.95} = \mathbf{3.4569}$

(c) $\sqrt[3]{8} = \mathbf{2}$

(d) $\sqrt[4]{12} = \mathbf{1.8612}$

(e) $\sqrt[5]{1500} = \mathbf{4.3174}$

2.   Find the value of X to four places by using the guessing game.

(a) $2^X = 12$

   $\mathbf{X = 3.5850}$

(b) $12^X = 14.65$

   $\mathbf{X = 1.0803}$

(c) $X^3 = 20$

   $\mathbf{X = 2.7144}$

(d) $X^2 = 11.95$

   $\mathbf{X = 3.4569}$

(e) $6.3^{(2x)} = 45$

   $\mathbf{X = 1.0341}$

(f) $10^{(X+1)} = 40$

X = .6021

3.   Steve earned 8% interest on his $3,000 investment yearly.

   (a)   How many years will it take before Steve has $4,408?
   6 Years

   (b)   How many years will it take before Steve has $10,277.83?
   16 Years

   (c)   If Steve wanted to guarantee himself a $15,000 return in 10 years, and he still had an 8% interest rate, how much should he invest initially?
   About $7,000 or exactly $6,947.90

4.   Jillian had the following scores for 18 holes of golf at Chipamonk Country Club

Day	Score
1	118
2	113
3	115
4	110
5	112
6	107
7	109

   (a)   How many days will it take before Jillian breaks 100?
   12 days

   (b)   Par for 18 hole at Chipamonk is 72. How many days will it take before Jillian reaches par?
   On the 30th day

   (c)   What is the pattern you see with Jillian's scores
   Jillian's scores decrease by three golf strokes every 2 days.

Name _____   Date _____

Supplement No. 5-3

The Guessing Game

1. Find the following roots to 4 places by using the guessing game.

    (a) $\sqrt{7} =$

    (b) $\sqrt{11.95} =$

    (c) $\sqrt[3]{8} =$

    (d) $\sqrt[4]{12} =$

    (e) $\sqrt[5]{1500} =$

2. Find the value of X to four places by using the guessing game.

    (a) $2^X = 12$

    (b) $12^X = 14.65$

    (c) $X^3 = 20$

    (d) $X^2 = 11.95$

    (e) $6.3^{(2x)} = 45$

    (f) $10^{(X + 1)} = 40$

MATH *Connections*® 1
A Secondary Mathematics Core Curriculum   © 2000 MATHconx, LLC

3.	Steve earned 8% interest on his $3,000 investment yearly.

(a)	How many years will it take before Steve has $4,408?

(b)	How many years will it take before Steve has $10,277.83?

(c)	If Steve wanted to guarantee himself a $15,000 return in 10 years, and he still had an 8% interest rate, how much should he invest initially?

4.	Jillian had the following scores for 18 holes of golf at Chipamonk Country Club

Day	Score
1	118
2	113
3	115
4	110
5	112
6	107
7	109

(a)	How many days will it take before Jillian breaks 100?

(b)	Par for 18 hole at Chipamonk is 72. How many days will it take before Jillian reaches par?

(c)	What is the pattern you see with Jillian's scores

Supplement No. 5-4

Finding Intersection Points
(Suitable for use with Section 5.3 of Year 1)

To the Teacher: *This Supplement gives students practice with graphing and finding intersection points.*

1. For each of the following problems, use a graph to determine the coordinates of the point where each pair of lines intersect.

    (a) $y = -x$ **(1, 1)**
         $y = 3x - 4$

    (b) $y = 4x + 2$ $\left(\dfrac{2}{3}, \dfrac{14}{3}\right)$
         $y - 4 = x$

    (c) $y - 7 = x$ **(-2.5, 4.5)**
         $\dfrac{1}{5}x + y = 4$

    (d) $y - 2x = 7$ **No Solution**
         $2x - y = 3$

    (e) $3.5x - y = -5$ **(3, 15.5)**
         $4.5x - y = -2$

    (f) $2x - y = -8$ $\left(-\dfrac{28}{3}, -\dfrac{32}{3}\right)$
         $x - 2y = 12$

    (g) $y - 4x = 9$ **Just one line. (Infinite Solutions)**
         $y - 9 = 4x$

    (h) $x + 2y = -8$ **No Solution**
         $2y - 6 = -x$

2. Mother's Day is near and there are two local flower shops selling flowers. Flo's Flower Shop charges $1.75 per rose plus a delivery fee of $2.50. Fred's Flower Shop charges $2.25 per rose with free delivery.

    (a) Write an equation to find the total cost, T dollars, of having roses, r, delivered from Flo's Flower Shop.

    $$T = 1.75r + 2.50$$

(b)   Write an equation to find the total cost, T dollars, of having roses, r, delivered from Fred's Flower Shop.

$$T = 2.25r$$

(c)   If you were going to send your mother a dozen roses, from which shop would you buy? Why?

Flo's: T = 1.75(12) + 2.5 = $23.50
Fred's: T = 2.25(12) = $27.00
The cost of a dozen roses is less expensive at Flo's.

(d)   How many roses can you have sent to your mother so that the cost is the same no matter which shop provided the roses? What will this cost be?

1.75r + 2.5 = 2.25r
r = 5 roses for $11.50
This can also be done using the graph or table features of the graphing calculator.

3.   Ned Tendo open a video store in town. Ned has a membership package that charges $0.60 per video game rental plus a one-time membership fee of $7.75. Mario's video Store down the street charges $1.20 per video game rental plus a one-time membership fee of $1.30.

(a)   How much would it cost you to rent three video games at each video store for the first time?

Ned's: .60(3) + 7.75 = 9.33
Mario's: 1.20(3) + 1.30 = $4.90

(b)   If you were going to rent seven video games for a party, which video store would you rent from for the first time?

Ned's: .60(7) + 7.75 = $11.95
Mario's: 1.2(7) + 1.30 = $9.70

(c)   How many video games would you have to rent for the first time so that the cost was the same at both stores? What would this cost be?
The point of intersection is (10.75, 14.20) but you can't rent 10.75 games. It would cost less to rent 10 games from Mario and 11 games from Ned's.

Name _____   Date _____

Finding Intersection Points

1. For each of the following problems, use a graph to determine the coordinates of the point where each pair of lines intersect.

   (a) $y = -x$
        $y = 3x - 4$

   (b) $y = 4x + 2$
        $y - 4 = x$

   (c) $y - 7 = x$
        $\frac{1}{5}x + y = 4$

   (d) $y - 2x = 7$
        $2x - y = 3$

   (e) $3.5x - y = -5$
        $4.5x - y = -2$

   (f) $2x - y = -8$
        $x - 2y = 12$

   (g) $y - 4x = 9$
        $y - 9 = 4x$

   (h) $x + 2y = -8$
        $2y - 6 = -x$

2. Mother's Day is near and there are two local flower shops selling flowers. Flo's Flower Shop charges $1.75 per rose plus a delivery fee of $2.50. Fred's Flower Shop charges $2.25 per rose with free delivery.

   (a) Write an equation to find the total cost, T dollars, of having roses, r, delivered from Flo's Flower Shop.

(b)   Write an equation to find the total cost, T dollars, of having roses, r, delivered from Fred's Flower Shop.

(c)   If you were going to send your mother a dozen roses, from which shop would you buy? Why?

(d)   How many roses can you have sent to your mother so that the cost is the same no matter which shop provided the roses? What will this cost be?

3.   Ned Tendo open a video store in town. Ned has a membership package that charges $0.60 per video game rental plus a one-time membership fee of $7.75. Mario's video Store down the street charges $1.20 per video game rental plus a one-time membership fee of $1.30.

(a)   How much would it cost you to rent three video games at each video store for the first time?

(b)   If you were going to rent seven video games for a party, which video store would you rent from for the first time?

(c)   How many video games would you have to rent for the first time so that the cost was the same at both stores? What would this cost be?

Supplement No. 5-5

Missing Points
(Suitable for use with Section 5.4 of Year 1)

To the Teacher: *The student will find missing value and complete the table and graph points. Neatness is important in this Supplement.*

1. Complete the table so that each point is on its graph: $4m + 3n = 96$ or $5c - 3d = 90$

m	n	c	d
0	32	0	-30
24	0	18	0
$\frac{1}{4}$	$32\frac{2}{3}$	0.2	-29.67
$23\frac{3}{4}$	$\frac{1}{3}$	18.24	0.4
$\frac{1}{2}$	$31\frac{1}{3}$	0.6	-29
$23\frac{1}{2}$	$\frac{2}{3}$	18.48	0.8
$\frac{3}{4}$	31	1.2	-28
$23\frac{1}{4}$	1	18.84	1.4
1	$30\frac{2}{3}$	1.6	-27.33
23	$1\frac{1}{3}$	19.08	1.8
2	$30\frac{1}{3}$	2.4	-26
$22\frac{3}{4}$	$2\frac{1}{3}$	19.56	2.6
$3\frac{3}{4}$	$29\frac{1}{3}$	2.8	-25.33
$22\frac{1}{4}$	4	19.92	3.2

2. Plot answers on the graph paper provided.

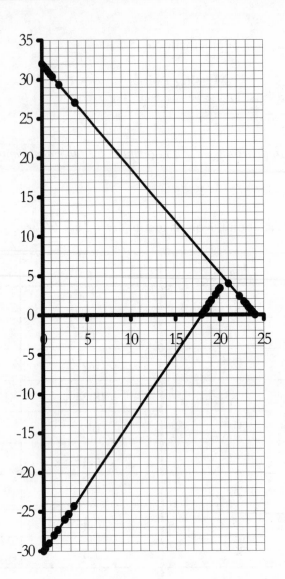

Supplement No. 5-5

Missing Points

1.   Complete the table so that each point is on its graph: $4m + 3n = 96$ or $5c - 3d = 90$

m	n	c	d
0		0	
	0		0
$\frac{1}{4}$		0.2	
	$\frac{1}{3}$		0.4
$\frac{1}{2}$		0.6	
	$\frac{2}{3}$		0.8
$\frac{3}{4}$		1.2	
	1		1.4
1		1.6	
	$1\frac{1}{3}$		1.8
2		2.4	
	$2\frac{1}{3}$		2.6
$3\frac{3}{4}$		2.8	
	4		3.2

2.   Plot answers on the graph paper provided.

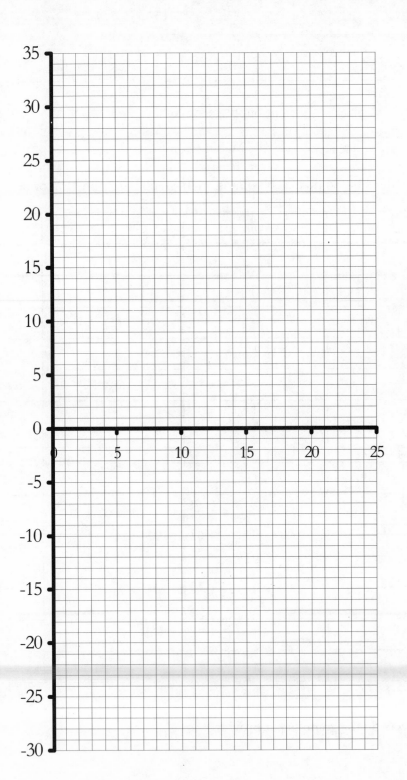

Supplement No. 5-6

Solving Equations
(Suitable for use with Section 5.4 of Year 1)

To the Teacher: *This Supplement gives students experience changing formulas to the slope intercept form.*

1. Solve each equation for y. (Slope-intercept form: $y = mx + b$)

   (a) $2x + y = 6$

   $y = -2x + 6$

   (b) $3x - 2y = 12$

   $y = \dfrac{3}{2}x - 6$

   (c) $x + 2y = 5$

   $y = \dfrac{-1}{2}x + \dfrac{5}{2}$

   (d) $2(x - y) = 10$

   $y = x - 5$

   (e) $3y = 2x + 12$

   $y = \dfrac{2}{3}x + 4$

   (f) $2y - 5x = -13$

   $y = \dfrac{5}{2}x - \dfrac{13}{2}$

   (g) $-y - 8 = x + 4$

   $y = -x - 12$

   (h) $2(2x - y) = -\dfrac{1}{2}(6x + 10y)$

   $y = \dfrac{-7}{3x}$

2. Give two reasons why we would like to manipulate equations to slope-intercept form.

   Reason 1: The graphing calculator only allows an equation in Y= form to be keyed in.

Reason 2: The slope and the Y-intercept are important number in real life problems and in graphing equations by hand.

3. Explain why the following two equations are not linear.

    (a) $y = 3x^2$
    (b) $y = \sqrt{x}$

    These equations are in radical form or have an exponent other than 1 with the variable.

4. Why is it important to know that some equations have curved lines as opposed to straight lines? Think in terms of intersecting lines.

    Curved lines may intersect more than once.

5. Find the intersection point(s) for each system of equations with your calculator. You may have to adjust the window on some problems.

    (a) $y = 2x + 3$
         $y = x + 4$

         (1, 5)

    (b) $y = 3x - 5$
         $y = \frac{1}{2}x + 10$

         (6, 13)

    (c) $y = 2(x - 3)$
         $y = -2 - 2x$

         (1, -4)

    (d) $y = 4x - 5$
         $y = -6x$

         $\left(\frac{1}{2}, -3\right)$

    (e) $y = 2x^2$
         $y = x + 1$

         (1, 2) and $\left(-\frac{1}{2}, \frac{1}{2}\right)$

    (f) $y = 3x^2$
         $y = -3x^2 + 5$

         (.92, 2.5) and (-.92, 2.5)

6. Write the equations of a line in slope-intercept form that does not intersect with the given line.

Answers will vary

(a) $y = 3x - 1$

y = 3x + 2

(b) $y = -2x + 5$

y = -2x + 8

(c) $y = .7x - 1$

y = .7x + 6

(d) $y = -\frac{2}{3}x + 5$

y = $\frac{-2}{3}$x − 5

(e) $2x + y = 6$

y = -2x + 1

(f) $3x - 7y = 14$

y = $\frac{3}{7}$x

(g) $2y - x = -4$

y = $\frac{1}{2}$x + 12

(h) $3x - 2y = 2(x + y)$

y = $\frac{1}{4}$x + 8

(i) $y = 5$

y = 2

7. Consider $y = 6x^2$. For parts (a) – (f), write the equation of a line that meets the following conditions.

Answers will vary.

(a) A curved line that intersects exactly twice

y = -6x^2 + 3

(b)   A curved line that intersects exactly once

$y = -2x^2$

(c)   A curved line that never intersects

$y = -3x^2 + 5$

(d)   A straight line that intersects exactly twice

$y = -3x + 4$

(e)   A straight line that intersects exactly once

$y = \frac{1}{8}x$

(f)   A straight line that never intersects

$y = \frac{1}{5}x - 6$

Supplement No. 5-6

Solving Equations

1. Solve each equation for y. (Slope-intercept form: $y = mx + b$)

    (a) $2x + y = 6$

    (b) $3x - 2y = 12$

    (c) $x + 2y = 5$

    (d) $2(x - y) = 10$

    (e) $3y = 2x + 12$

    (f) $2y - 5x = -13$

    (g) $-y - 8 = x + 4$

    (h) $2(2x - y) = -\frac{1}{2}(6x + 10y)$

2. Give two reasons why we would like to manipulate equations to slope-intercept form.

3. Explain why the following two equations are not linear.

    (a) $y = 3x^2$
    (b) $y = \sqrt{x}$

4. Why is it important to know that some equations have curved lines as opposed to straight lines? Think in terms of intersecting lines.

5. Find the intersection point(s) for each system of equations with your calculator. You may have to adjust the window on some problems.

(a) $y = 2x + 3$
$y = x + 4$

(b) $y = 3x - 5$
$y = \frac{1}{2}x + 10$

(c) $y = 2(x - 3)$
$\mathbf{y} = -2 - 2x$

(d) $y = 4x - 5$
$y = -6x$

(e) $y = 2x^2$
$y = x + 1$

(f) $y = 3x^2$
$y = -3x^2 + 5$

6. Write the equations of a line in slope-intercept form that does not intersect with the given line.
Answers will vary

(a) $y = 3x - 1$

(b) $y = -2x + 5$

(c) $y = .7x - 1$

(d) $y = -\frac{2}{3}x + 5$

(e) $2x + y = 6$

(f) $3x - 7y = 14$

(g) $2y - x = -4$

(h) $3x - 2y = 2(x + y)$

(i) $y = 5$

7. Consider $y = 6x^2$. For parts (a) – (f), write the equation of a line that meets the following conditions.

(a) A curved line that intersects exactly twice

(b) A curved line that intersects exactly once

(c) A curved line that never intersects

(d) A straight line that intersects exactly twice

(e) A straight line that intersects exactly once

(f) A straight line that never intersects

Supplement No. 5-7

Graphing Lines
(Suitable for use with Section 5.6 of Year 1)

To the Teacher: *This exercise provides students with the opportunity to write equations in the $y = mx + b$ for and identify x and y intercepts.*

Fill in the chart and graph the equations on graph paper.

	Equations	Solve for y	Slope m =	x	y	y - intercept	x - intercept
1.	$3x + 2y = 12$	$y = \left(\frac{-3}{2}\right)x + 6$	$m = \frac{-3}{2}$	-2	9	0, 6	4, 0
				-1	7.5		
				0	6		
				1	4.5		
				2	3		
2.	$4x + 5y = 20$	$y = \frac{4x}{5} + 4$	$m = \frac{4}{5}$	-2	2.4	0, 4	-5, 0
				-1	3.2		
				0	4		
				1	4.8		
				2	5.6		
3.	$-2x - 8y = 24$	$y = \frac{-x}{4} - 3$	$m = \frac{-1}{4}$	-2	-2.5	0, -3	-12, 0
				-1	-2.75		
				0	-3		
				1	-3.25		
				2	-3.5		
4.	$x - y = 7$	$y = x - 7$	$m = 1$	-2	-9	0, -7	7, 0
				-1	-8		
				0	-7		
				1	-6		
				2	-5		

Name _____   Date _____

Supplement No. 5-7

Graphing Lines

Fill in the chart and graph the equations on graph paper.

	Equations	Solve for y	Slope m =	x	y	y - intercept	x - intercept
1.	$3x + 2y = 12$			-2			
				-1			
				0			
				1			
				2			
2.	$4x + 5y = 20$			-2			
				-1			
				0			
				1			
				2			
3.	$-2x - 8y = 24$			-2			
				-1			
				0			
				1			
				2			
4.	$x - y = 7$			-2			
				-1			
				0			
				1			
				2			

Supplement No. 5-8

Graphing Real Life Situations
(Suitable for use with Section 5.6 of Year 1)

To the Teacher *Using real life information, this Supplement provides students with experiences in graphing equations on the graphing calculator. Note: Gasoline prices vary throughout the United States.*

Today's gasoline prices are always changing. Most gasoline stations have three different prices based on the octane level of the gasoline. In July 1998 a major gasoline company charged $1.25 a gallon for the regular gas, $1.29 a gallon for the premium gas and $1.35 a gallon for the super premium gas. Use your graphing calculator and construct equations to solve each of the following situations.

1.   How much does it cost to fill a sixteen-gallon tank with:

   (a)   Regular gas?   **$20.00**

   (b)   Premium gas?   **$20.64**

   (c)   Super Premium gas?   **$21.60**

   (d)   How much more does it cost to fill the tank with Super Premium than Regular?

       $21.60 – $20.00 = $1.60

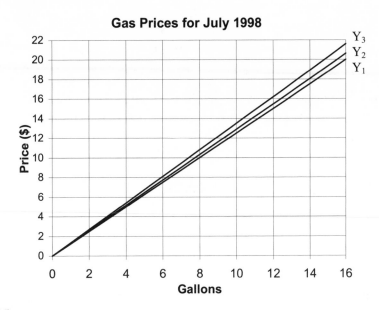

Gas Prices for July 1998

$Y_1 = 1.25x$
$Y_2 = 1.29x$
$Y_3 = 1.35x$

2.   If a driver fills his car once a week how much does he pay for 4 weeks of:

(a)   Regular gas?   **$80.00**

(b)   Premium gas?   **$82.56**

(c)   Super Premium gas?   **$86.40**

(d)   How much more does it cost to buy 4 weeks of Super Premium than Regular?

$86.40 − $80.00 = $6.40

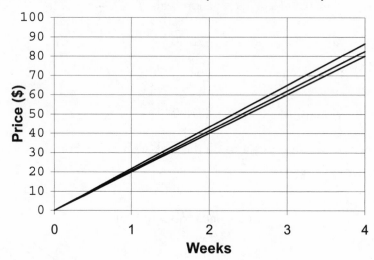

Gas Costs Per Week (16 Gallon Tank)

If gasoline prices go back up to their 1990's highs, the charge would be $1.46 a gallon for regular gas, $1.69 a gallon for premium gas and $1.80 a gallon for super premium.

3.   How much does it cost to fill a 16-gallon tank with:

(a)   Regular gas?   **$23.36**

(b)   Premium gas?   **$27.04**

(c)   Super Premium?   **$28.80**

(d)   How much more for Super Premium than Regular?

$28.80 - $23.36 = $5.44

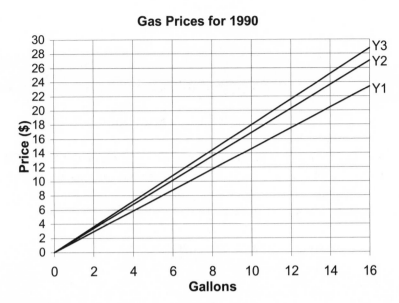

Gas Prices for 1990

$Y_1 = 1.46x$
$Y_2 = 1.69x$
$Y_3 = 1.80x$

4. What is the difference in the cost of one year of gas (once a week for 52 weeks) at the 1990 prices and the July 1998 prices.

    (a) Regular gas? $1214.72 − $1040 = $174.72

    (b) Premium gas? $1406 − $1073.28 = $332.80

    (c) Super Premium gas? $1497.60 − $1123 = $374.40

5. Graph the equations for the least expensive regular price (July 1998) and the cost of the highest super premium gas (1990's high). What is the major difference?

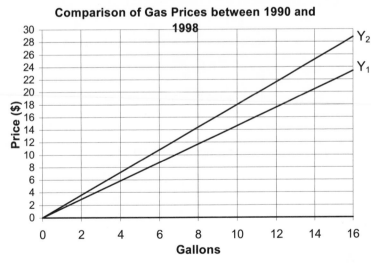

Comparison of Gas Prices between 1990 and 1998

The slope of y = 1.80x is much steeper than y = 1.25x.

Name _____   Date _____

Graphing Real Life Situations

Today's gasoline prices are always changing. Most gasoline stations have three different prices based on the octane level of the gasoline. In July 1998 a major gasoline company charged $1.25 a gallon for the regular gas, $1.29 a gallon for the premium gas and $1.35 a gallon for the super premium gas. Use your graphing calculator and construct equations to solve the following:

1. How much does it cost to fill a sixteen-gallon tank with:

   (a) Regular gas?

   (b) Premium gas?

   (c) Super Premium gas?

   (d) How much more for Super Premium than Regular?

2. If a driver fills his car once a week how much does he pay for 4 weeks of:

   (a) Regular gas?

   (b) Premium gas?

   (c) Super Premium gas?

   (d) How much more for 4 weeks of Super Premium than Regular?

If gasoline prices go back up to their 1990's highs, the charge would be $1.46 a gallon for regular gas, $1.69 a gallon for premium gas and $1.80 a gallon for super premium.

3. How much does it cost to fill a sixteen-gallon tank with:

   (a) Regular gas?

   (b) Premium gas?

   (c) Super Premium?

   (d) How much more for Super Premium than Regular?

4.   What is the difference in the cost of gas (once a week for 52 weeks) at the 1990 prices and the July 1998 prices.

(a)   Regular gas?

(b)   Premium gas?

(c)   Super Premium gas?

5.   Graph the equations for the least expensive regular price (July 1998) and the cost of the highest super premium gas (1990). What is the major difference?

Supplement No. 5-9

Solving Systems of Equations
(Suitable for use with Section 5.7 of Year 1)

To the Teacher: *This Supplement gives the student experience with solving systems of equations two ways, graphically and algebraically.*

Solve each problem algebraically and then graphically.

1. $4x - 3y = 17$
     $2x + y = 11$

   (a)   Algebraically   (b)   Graphically

Multiply by (-2) $4x - 3y = 17$
$4x - 3y = 17$ $2x + y = 11$
$-4x - 2y = -22$
   $-5y = -5$ **Use a Graphing Calculator to graph**
      $y = 1$ $Y_1 = \dfrac{4x}{3} - \dfrac{17}{3}$

$4x - 3(1) = 17$ $Y_2 = -2x + 11$
   $4x = 20$
    $x = 5$

      (5, 1)   **Point of Intersection (5, 1)**

2. $-3x + 2y = -1$
      $x - 3y = 5$

   (a)   Algebraically   (b)   Graphically

   $-3x + 2y = -1$ $-3x + 2y = -1$
   $3x - 9y = 15$ $x - 3y = 5$
     $-7y = 14$
       $y = -2$ **Use a Graphing Calculator to graph**

   $-3x + 2(-2) = -1$

       $-3x = 3$ $Y_1 = \dfrac{3x}{2} - \dfrac{1}{2}$
        $x = -1$ $Y_2 = \dfrac{x}{3} - \dfrac{5}{3}$

      (-1, -2)   **Point of Intersection (-1, -2)**

3. $x - 4y = 15$
       $-x + 2y = -9$

     (a)   Algebraically   (b)   Graphically

$x - 4y = 15$	$x - 4y = 15$
$\underline{-x + 2y = -9}$	$-x + 2y = -9$
$-2y = 6$	
$y = -3$	**Use a graphing calculator to graph**
$x - 4(-3) = 15$	$Y_1 = \dfrac{x}{4} - \dfrac{15}{4}$
$x = 3$	$Y_2 = \dfrac{x}{2} - \dfrac{9}{2}$
$(3, -3)$	**Point of Intersection $(3, -3)$**

4. $-3x + 2y = 12$
       $-x + 5y = 4$

     (a)   Algebraically   (b)   Graphically

$3x + 2y = 12$	$-3x + 2y = 12$
$3x - 15y = -12$	$-x + 5y = 4$
$-13y = 0$	
$y = 0$	**Use the graphing calculator to graph**
$-3x + 2(0) = 12$	$Y_1 = \dfrac{3x}{2} + 6$
$-3x = 12$	$Y_2 = \dfrac{x}{5} + \dfrac{4}{5}$
$x = -4$	
$(-4, 0)$	**Point of Intersection $(-4, 0)$**

Name _____   Date _____

Supplement No. 5-9

Solving Systems of Equations

Solve each problem algebraically and then graphically.

1. $4x - 3y = 17$
    $2x + y = 11$

    (a)   Algebraically   (b)   Graphically

2. $-3x + 2y = -1$
    $x - 3y = 5$

    (a)   Algebraically   (b)   Graphically

3. $x - 4y = 15$
    $-x + 2y = -9$

    (a)   Algebraically   (b)   Graphically

4. $-3x + 2y = 12$
    $-x + 5y = 4$

    (a)   Algebraically   (b)   Graphically

Supplement No. 6-1

It All Depends - Classic McDonald's Meal
(Suitable for use with Section 6.1 of Year 1)

To the Teacher: *This supplement gives the students experience with functions.*

1. In the 1960's, the most popular (classic) meal cost $0.15 for a hamburger, $0.10 for french-fries, and $0.12 for a milk shake.

   (a) Write a formula for the cost of a classic McDonald's meal at that time.

   c(x) = $0.15x + $0.10x + $0.12x
   c(x) = $0.37x

   (b) In the 1960's, how much would it cost a family of 5 to buy the classic McDonald's meal for each person?

   c(x) = $0.37x
   c(5) = $0.37(5)
   c(5) = $1.85

   (c) In the 1960's, 35 students in a school bus went to a McDonald's and each ordered a classic McDonald's meal. How much was the total bill?

   c(x) = $0.37x
   c(35) = $0.37(35)
   c(35) = $12.95

2. In 1998, the prices for the classic McDonald's meal were: $0.85 for a hamburger, $0.89 for french-fries and $1.19 for a milk shake.

   (a) Write a formula for the cost of a 1998 classic McDonald's meal.

   c(x) = $0.85x + $0.89x + $1.19x
   c(x) = $2.93x

   (b) In 1998, how much did it cost a family of 5 to each buy the classic McDonald's meal

   c(x) = $2.93x
   c(5) = $14.65

   (c) In 1998, 35 students on a school bus each ordered a classic McDonald's meal. How much was the total bill? (omit sales tax)

   c(x) = $2.93x
   c(35) = $102.55

3.   Use the table function of your graphing calculator to determine the difference in prices for ordering the classic McDonald's meal in the 1960' and 1998 for:

(a)   A family of 5

(b)   A busload of 35

(c)   A school of 420

(d)   A town of 30,000

	# X	1960 Y1	1998 Y2	Difference Y3
(a)	5	$1.85	$14.65	$12.60
(b)	35	$12.95	$102.55	$89.60
(c)	420	$155.40	$1,230.60	$1,075.20
(d)	30,000	$11,100.00	$87,900.00	$76,800.00

Name _____   Date _____

It All Depends - Classic McDonald's Meal

1.   In the 1960's, the most popular (classic) meal cost $0.15 for a hamburger, $0.10 for french-fries, and $0.12 for a milk shake.

   (a)   Write a formula for the cost of a classic McDonald's meal at that time.

   (b)   In the 1960's, how much would it cost a family of 5 to buy the classic McDonald's meal for each person?

   (c)   In the 1960's, 35 students in a school bus went to a McDonald's and each ordered a classic McDonald's meal. How much was the total bill?

2.   In 1998, the prices for the classic McDonald's meal were: $0.85 for a hamburger, $0.89 for french-fries and $1.19 for a milk shake.

   (a)   Write a formula for the cost of a 1998 classic McDonald's meal.

   (b)   In 1998, how much did it cost a family of 5 to each buy the classic McDonald's meal

   (c)   In 1998, 35 students on a school bus each ordered a classic McDonald's meal. How much was the total bill? (omit sales tax)

3. Use the table function of your graphing calculator to determine the difference in prices for ordering the classic McDonald's meal in the 1960' and 1998 for:

   (a) A family of 5

   (b) A busload of 35

   (c) A school of 420

   (d) A town of 30,000

Supplement No. 6-2

It All Depends – Buying Newspapers
(Suitable for use with Section 6.1 of Year 1)

To the Teacher: *This supplement gives students the opportunity to create a function and then use it in several different situations.*

1. The cost of a daily newspaper in a mid-western state is $0.50 plus a sales tax of 6%.

   (a) Find the costs of the following and put the results in a table.

      (1) 7 daily newspapers
      $y_1 = .53x$
      $d(7) = \$3.71$

      (2) 15 daily newspapers
      $d(15) = \$7.95$

      (3) 313 daily newspaper
      $d(313) = \$165.89$

   Do you see a pattern? Explain.

   The costs increase proportionately by .53.

   (b) Write an equation for buying x-number of newspapers including the tax. Verify your results in part (a) using your equation.

   $d(x) = .50x + .06(.50x)$
   $d(x) = .53x$

   (c) If the cost of a Sunday newspaper is $2.00 plus a sales tax of 6 %, how much would you spend in a year on Sunday newspapers (52 weeks in a year).

   $s(x) = 2.00x + .06(2.00x)$
   $s(x) = 2.12x$
   $110.24

   (d) What is the cost of buying a newspaper every day of the week, for the entire year? (Use the information in parts (a), (b) and (c))

   $d(313) = \$165.89$
   $\$165.89 + \$110.24 = \$276.13$
   *Note: 365-52=313, except for leap year.

2. The same mid-western state no longer taxes newspapers.

   (a) Write a new equation for buying x-number of daily and Sunday newspapers in the state.

   $d(x) = .50x$
   $s(x) = 2.00x$

(b)   What is the cost of buying a newspaper every day of the week, for the entire year without tax?
d(313) = 156.50
s(52) = 104.00
$260.50

(c)   How much did we save without the sales tax for one year?
$276.13 – 260.50 = $15.63

(d)   If the state's leading newspaper sold 425,000 daily newspapers, and 550,000 Sunday newspapers in a week, how much money did the state lose from ending the sales tax?

Daily: $225,250 – $212,500 = $12,750
Sunday: $1,166,000 - $1,110,000 = $56,000
State lost $68,750

Name _____   Date _____

It All Depends – Buying Newspapers

1.   The cost of a daily newspaper in a mid-western state is $0.50 plus a sales tax of 6%.

   (a)   Find the costs of the following and put the results in a table.

       (1)   7 daily newspapers

       (2)   15 daily newspapers

       (3)   313 daily newspaper

   Do you see a pattern? Explain.

   (b)   Write an equation for buying x-number of newspapers including the tax. Verify your
          results in part (a) using your equation.

   (c)   If the cost of a Sunday newspaper is $2.00 plus a sales tax of 6%, how much would you
          spend in a year on Sunday newspapers (52 weeks in a year).

   (d)   What is the cost of buying a newspaper every day of the week, for the entire year? (Use
          the information in parts (a), (b) and (c))

2.   The same mid-western state no longer taxes newspapers.

   (a)   Write a new equation for buying x-number of newspapers in the state.

   (b)   What is the cost of buying a newspaper everyday of the week, for the entire year?

   (c)   How much did we save without the sales tax for one year?

   (d)   If the state's leading newspaper sold 425,000 daily newspapers, and 550,000 Sunday newspapers, how much money did the State lose from ending the sales tax?

Supplement No. 6-3

A Simple Sequence Study
(Suitable for use with Section 6.2 of Year 1)

This exercise requires the student to find missing terms in a sequence, define the sequence using recursive and rule formulae, and determine a specific term using the rule of the sequence.

1.   A sequence S begins as :   3, 9, 15, 2, 27, . . .

   (a)   What are the next two terms ?   How did you find them ?

   33, 39   Add 6 to previous term.

   (b)   Define this sequence recursively.

   $$S_n = S_{n-1} + 6$$

   (c)   Write a formula for S(n):

   S(n) = 6n - 3

   (d)   Find S(120)

   S(120) = 591

2.   A sequence S begins as :   39, 35, 31, 27, 23, . . .

   (a)   What are the next five terms ?   How did you find them ?

   19 , 15 , 11 , 7 , 3   **Subtract 4   from previous term**

   (b)   Define this sequence recursively.
   $$S_n = S_{n-1} - 4$$

   (c)   Write a formula for S(n):

   S(n) = -4n + 23

   (d)   Find S(15)

   S(15) = -37

3.   A sequence S begins as :   23, 18, 13, 8, 3, . . .

    (a)   What are the next three terms ?   How did you find them ?

        -2, -7, -12   **Subtract 5 from previous term**

    (b)   Define this sequence recursively.

        $S_n = S_{n-1} - 5$

    (c)   Write a formula for S(n):

        S(n) = -5n + 28

    (d)   Find S(100)

        S(100) = -472

4.   A sequence S begins as :   12, 12.75, 13.5, 14.2 , 15, . . .

    (a)   What are the next four terms ?   How did you find them ?

        15.75, 16.5, 17.25, 18   **Add 0.75 to previous term**

    (b)   Define this sequence recursively.

        $S_n = S_{n-1} + 0.75$

    (c)   Write a formula for S(n):

        S(n) = 0.75n + 15

    (d)   Find S(25)

        S(25) = 33.75

5.   A sequence S begins as :   2.6, 2.2, 1.8 , 1.4, 1.0, . . .

    (a)   What are the next four terms ?   How did you find them ?

        0.6, 0.2, -0.2, -0.6   **Subtract 0.4 from previous term**

    (b)   Define this sequence recursively.

        $S_n = S_{n-1} - 0.4$

    (c)   Write a formula for S(n):

        S(n) = -0.4n + 3

    (d)   Find S(40)

        S(40) = -13

Name _____   Date _____

Sequences

1. A sequence S begins as :   3, 9, 15, 21, 27, . . .

    (a)   What are the next two terms ? How did you find them ?

    (b)   Define this sequence recursively.

    (c)   Write a formula for S(n):

    (d)   Find S(120)

2. A sequence S begins as :   39, 35, 31, 27, 23, . . .

    (a)   What are the next five terms ? How did you find them ?

    (b)   Define this sequence recursively.

    (c)   Write a formula for S(n):

(d)   Find S(15)

3.   A sequence S begins as :   23, 18, 13, 8, 3, . . .

(a)   What are the next three terms ?   How did you find them ?

(b)   Define this sequence recursively.

(c)   Write a formula for S(n):

(d)   Find S(100)

4.   A sequence S begins as :   12, 12.75, 13.5, 14.25, 15, . . .

(a)   What are the next four terms ?   How did you find them ?

(b)   Define this sequence recursively.

(c)   Write a formula for S(n):

(d)   Find S(25)

5.   A sequence S begins as :   2.6, 2.2, 1.8, 1.4, 1.0, . . .

(a)   What are the next four terms ?   How did you find them ?

(b)   Define this sequence recursively.

(c)   Write a formula for S(n):

(d)   Find S(40)

Supplement No. 6-4

Telephone Problem
(Suitable for use with Section 6.3 of Year 1)

To the Teacher: *The student needs to analyze the requirements carefully and appropriately graph the results.*

1. In 1998 one of the New England states charged the following rate for a local phone call from a pay phone. The first five minutes were $0.25 and each additional five minutes cost $0.50. Complete the chart:

Time	Number of Minutes Since 8:00	Cost in dollars
8:00 - 8:05	$0 < n \leq 5$	.25
8:05 - 8:10	$5 < n \leq 10$	.50 +.25 = .75
8:10 - 8:15	$10 < n \leq 15$	1.00 +.25 = 1.25
8:15 - 8:20	$15 < n \leq 20$	1.50 + .25 = 1.75
8:20 - 8:25	$20 < n \leq 25$	2.00 + .25 = 2.25
8:25 - 8:30	$25 < n \leq 30$	2.50 + .25 = 2.75
8:30 - 8:35	$30 < n \leq 35$	3.00 + .25 = 3.25
8:35 - 8:40	$35 < n \leq 40$	3.50 + .25 = 3.75
8:40 - 8:45	$40 < n \leq 45$	4.00 + .25 = 4.25
8:45 - 8:50	$45 < n \leq 50$	4.50 + .25 = 4.75
8:50 - 8:55	$50 < n \leq 55$	5.00 + .25 = 5.25
8:55 - 9:00	$55 < n \leq 60$	5.50 + .25 = 5.75

2. Now make a step function graph.

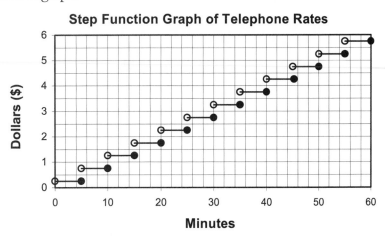

Step Function Graph of Telephone Rates

Supplement No. 6-4

Telephone Problem

1. In 1998 one of the New England states charged the following rate for a local phone call from a pay phone. The first five minutes were $0.25 and each additional five minutes cost $0.50. Complete the chart:

Time	Number of Minutes Since 8:00	Cost in dollars
8:00 - 8:05	$0 < n \leq 5$	.25
8:05 - 8:10	$5 < n \leq 10$	.50 + .25 = .75
8:10 - 8:15	$10 < n \leq 15$	1.00 + .25 = 1.25
8:15 - 8:20		
8:20 - 8:25		
8:25 - 8:30		
8:30 - 8:35		
8:35 - 8:40		
8:40 - 8:45		
8:45 - 8:50		
8:50 - 8:55		
8:55 - 9:00		

2. Now make a step function graph.

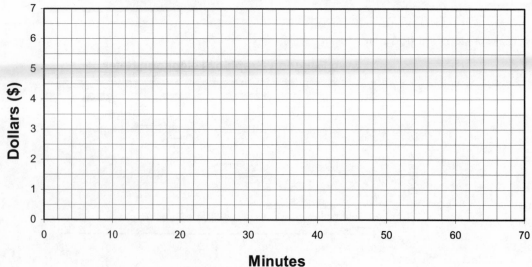

Step Function Graph of Telephone Rates

Supplement No. 6-5

Describing Functions with Algebra
(Suitable for use with Section 6.4 of Year 1)

To the Teacher: *This supplement gives the student more experience converting from miles to kilometers and vice versa.*

1. Every serious runner knows that a 10k (kilometer race) is approximately 6.2 miles.

    (a) Write a formula for miles in terms of kilometers.

    m(K) = .62k

    (b) Use your graphing calculator to determine how many miles in a:

    (1.) 5K race

    m(5) = 3.10 miles

    (2.) 20K race

    m(20) = 12.4 miles

    (3.) 50K race

    m(50) = 31 miles

    (c) Write a formula for kilometers in terms of miles.

    k(m) = 1.61m

    (d) Use your graphing calculator to determine how many kilometers are in the following problems:

    (1.) 10 miles

    k(10) = 16.10 kilometers

    (2.) 50 miles

    k(50) = 80.5 kilometers

    (3.) 100 miles

    k(100) = 161 kilometers

(e) While traveling in other countries, a traveler noticed that distances between cities are given in kilometers. How many miles apart are cities with the following distances between them?

   (1.) 32 kilometers

   m(32) = 19.84

   (2.) 77 kilometers

   m(77) = 47.74

   (3.) 125 kilometers

   m(125) = 77.50

   (4.) 161 kilometers

   m(161) = 99.82

Name _____   Date _____

Describing Functions with Algebra

1.   Every serious runner knows that a 10k (kilometer race) is approximately 6.2 miles.

(a)   Write a formula for miles in terms of kilometers.

(b)   Use your graphing calculator to determine how many miles in a:

    (1.)   5K race

    (2.)   20K race

    (3.)   50K race

(c)   Write a formula for kilometers in terms of miles.

(d)   Use your graphing calculator to determine how many kilometers are in the following problems:

    (1.)   10 miles

    (2.)   50 miles

    (3.)   100 miles

(e)   While traveling in other countries, a traveler noticed that distances between cities are given in kilometers. How many miles apart are cities with the following distances between them?

    (1.)   32 kilometers

    (2.)   77 kilometers

    (3.)   125 kilometers

    (4.)   161 kilometers

MATH *Connections*® 1
A Secondary Mathematics Core Curriculum   S-1   © 1999 MATHconx, LLC

Supplement No. 6-6

Metric Bike Tour
(Suitable for use with Section 6.4 of Year 1)

Teacher Commentary: *This exercise includes more experience with conversion of miles to kilometers, kilometers to miles, calculating average speed, and revisits mean, median and mode.*

In Europe, bike races are measured in kilometers. In the United States, the distances of these races are measured in miles. One way to convert these two units of measure is to use the formula:

$$m = \left(\frac{5}{8}\right)k$$

1.   If one stage of a bike race is 170km, how far is that in miles ?
     106.25 miles

2.   Another stage of the bike race is 204km. How far is that in miles ?
     131.25 miles

3.   Calculate the distance (in km) for a 36.25 mile stage of the race.
     58 km

The following is a chart for the distances of one such metric bike tour. Use the **list** feature of your graphing calculator to complete this chart:

STAGE	DISTANCE (KM)	DISTANCE (MILES)
1	180.5	**112.8**
2	205.5	**128.44**
3	169	**105.63**
4	**252**	157.5
5	**228.5**	142.81
6	**204.5**	127.81
7	**58**	36.25
8	190.5	**119.06**
9	210	**131.25**
10	196.5	**122.81**
11	170	**106.25**
12	**222**	138.75
13	**196**	122.5
14	**186.5**	116.56
15	189	**118.13**
16	204	**127.5**
17	149	**93.13**
18	218.5	**136.56**
19	**242**	151.25
20	**52**	32.5
21	**147.5**	92.19

4. Calculate the total distance of this bike tour in km , in miles.

    3871.5 km   2419.7 miles

5. Using your calculator, find the *mean* , *median* , and *mode* for this data.

    mean: 184.4km / 115.2mi   median: 196km / 122.5 mi   mode: none

6. How many laps around a 1/4 mile high school track would it take to complete the distance of this bike tour ?

    9678.8 laps

7. How many 26.2 mile running marathons would it take to equal the distance of this bike tour ?

    approx. 92

8   How many round-trips from Boston to New York (210 mi) would be needed to match the total distance of this bike tour ?

    approx. 11.5 round-trips

9. If stage 8 of this tour was completed by the top rider in approximately 4 hours 40 minutes, then calculate this rider's average speed in km/hr and in mi/hr.

    40.79 km/hr   25.49 mi/hr

10. If stage 9 of this tour took 5 hours 21 minutes for the top rider to complete, then calculate this rider's average speed in km/hr and in mi/hr.

    39.25 km/hr   24.53 mi/hr

Name _____   Date _____

Supplement No. 6-6

Metric Bike Tour

In Europe, bike races are measured in kilometers. In the United States, the distances of these races are measured in miles. One way to convert these two units of measure is to use the formula:

$$m = \left(\frac{5}{8}\right)k$$

1.   If one stage of a bike race is 170km, how far is that in miles ?

2.   Another stage of the bike race is 204km. How far is that in miles ?

3.   Calculate the distance (in km) for a 36.25 mile stage of the race.

The following is a chart for the distances of one such metric bike tour. Use the **list** feature of your graphing calculator to complete this chart.

STAGE	DISTANCE (KM)	DISTANCE (MILES)
1	180.5	_____
2	205.5	_____
3	169	_____
4	_____	157.5
5	_____	142.81
6	_____	127.81
7	_____	36.25
8	190.5	_____
9	210	_____
10	196.5	_____
11	170	
12	_____	138.75
13	_____	122.5
14	_____	116.56
15	189	_____
16	204	_____
17	149	_____
18	218.5	_____
19	_____	151.25
20	_____	32.5
21	_____	92.19

4. Calculate the total distance of this bike tour in kilometers and miles.

5. Using your calculator, find the *mean* , *median* , and *mode* for this data.

    mean: median: mode:

6. How many laps around a 1/4 mile high school track would it take to complete the distance of this bike tour ?

7. How many 26.2 mile running marathons would it take to equal the distance of this bike tour ?

8   How many round-trips from Boston to New York (210 mi) would be needed to match the total distance of this bike tour ?

9. If stage 8 of this tour was completed by the top rider in approximately 4 hours 40 minutes, then calculate this rider's average speed in km/hr and in mi/hr.

10. If stage 9 of this tour took 5 hours 21 minutes for the top rider to complete, then calculate this rider's average speed in km/hr and in mi/hr.

Supplement No. 6-7

Inflation and Appreciation
(Suitable for use with Section 6.5 of Year 1)

To the Teacher: *This exercise gives the student additional practice calculating various values in growth functions. Students will use reasoning powers to help determine the correct solution.*

1. In its initial year, 1950, the most expensive brand of sneakers cost $19.95. If the doubling period for this brand of sneakers is sixteen years, what will the cost be in 1998?

    1998 – 1950 = 48 years
    C = 19.95 (2)^3 = $159.60

2. In approximately what year was this brand of sneakers worth $100?

    Using the graphing calculator with the Table Function, graph $Y_1 = 19.95(2^{\frac{x}{16}})$.
    This brand of sneakers will be worth $100 in approximately 1987

3. If you are 18 in 1998, how old would you be when the price is $451.42?

    The price will be $451.42 in 72 years from 1950.
    1950 + 72 years = 2022
    2022 – 1998 = 24 years
    18 years + 24 years = 42 years old

4. (a) What will the cost be on the 81st anniversary of the sneaker?

    $Y_1 = 19.95(2^{81/16}) = \$ 666.66$
    (or use the Table Function)

    (b) Would you buy the sneaker then?

    Answers will vary

5. In college basketball the varsity players do not have to pay for their sneakers. In the year 2020, how much would it cost to supply a college basketball team with sneakers for 12 players?

    $Y_1 = 19.95(2^{\frac{x}{16}})$
    $= 19.95(2^{\frac{70}{16}}) = \$413.95$
    413.95 x 12 = $4967.40
    The table value is 413.95 for the year 2020; therefore, $413.95 x 12 = $4,967.40.

6.   A famous painting worth $7,000 appreciates (annual growth rate) 10.5% per year.

(a)   How much would the painting be worth in 5 years?

$7000(1.105)^5 = \$11,532.13$

(b)   in 15 years?

$7000(1.105)^{15} = \$31,299.13$

7.   In 1990 a baseball card was worth $8.00 and appreciates 3.75% per year. How much would the card be worth in the following years:

(a)   1998

$8(1.0375)^8 = \$10.74$

(b)   2011

$8(1.0375)^{21} = \$17.33$

8.   A 1966 classic Ford Mustang in 1998 was worth $7,500. If it appreciates at the rate of 6%, then how much would the Mustang be worth in the following years:

(a)   2003

$7500(1.06)^5 = \$10,036.69$

(b)   2012

$7500(1.06)^{14} = \$16,956.78$

9.   What item is worth more: a $0.75 stamp appreciating at the rate of 2% in 56 years or a $1.75 comic book appreciating at the rate of 1% in 25 years?

$0.75 Stamp: $.75(1.02)^{56} = \$2.27$
$1.75 Comic Book: $1.75(1.01)^{25} = \$2.24$
The stamp is worth more.

10.   In how many years will the value of the $0.75 stamp appreciating at the rate of 2% be the same as the $1.75 comic book appreciating at the rate of 1%?

$0.75 Stamp: $.75(1.02)^{86} = \$4.12$
$1.75 Comic Book $1.75(1.01)^{86} = \$4.12$
The values will be the same in 86 years.

11.   An autographed basketball is worth $65 and it appreciates at the rate of 2.25% a year. In how many years will it double in value?

$65(1.0225)^{31} = \$129.56$ $65(1.0225)^{32} = \$132.48$

It will double in value in approximately 31 years. (Rule of 72: $\dfrac{72}{2.25} = 32$)

Name _____   Date _____

Supplement No. 6-7

Inflation and Appreciation

1.   In its initial year, 1950, the most expensive brand of sneakers cost $19.95. If the doubling period for this brand of sneakers is sixteen years, what will the cost be in 1998?

2.   In approximately what year was this brand of sneakers worth $100?

3.   If you were 18 in 1998, how old would you be when the price is $451.42?

4.   (a)   What will the cost be on the 81st anniversary of the sneaker?

     (b)   Would you buy the sneaker then?

5.   In college basketball the varsity players do not have to pay for their sneakers. In the year 2020, how much would it cost to supply a college basketball team with sneakers for 12 players?

6. A famous painting worth $7,000 appreciates (annual growth rate) 10.5% per year.

   (a) How much would the painting be worth in 5 years?

   (b) in 15 years?

7. In 1990 a baseball card was worth $8.00 and appreciates 3.75% per year. How much would the card be worth in the following years:

   (a) 1998

   (b) 2011

8. A 1966 classic Ford Mustang in 1998 was worth $7,500. If it appreciates at the rate of 6%, then how much would the Mustang be worth in the following years:

   (a) 2003

   (b) 2012

9. What item is worth more: a $0.75 stamp appreciating at the rate of 2% in 56 years or a $1.75 comic book appreciating at the rate of 1% in 25 years?

10. In how many years will the value of the $0.75 stamp appreciate at the rate of 2% be the same as the $1.75 comic book appreciating at the rate of 1%?

11. An autographed basketball is worth $65 and it appreciates at the rate of 2.25% a year. In how many years will it double in value?

Supplement No. 6-8

Function Composition
(Suitable for use with Section 6.6 of Year 1)

To the Teacher: *This supplement gives the students additional experience in evaluating the graphic and numeric composition of functions.*

1. Refer to the graphs below to approximate each of the following quantities.
   Many of the following answers are approximate because one is reading values from a graph.

   (a)   f(2) = **2.5**

   (b)   g(f(2)) = **3.25**

   (c)   f(5) = **3**

   (d)   g(f(5)) = **3.5**

   (e)   g(3) = **3.5**

   (f)   f(g(3)) = **4.5**

   (g)   g(2) = **3**

   (h)   f(g(2)) = **3.75**

   (i)   Compare the answers from (b) and (h). Is the function composition commutative? Explain.

   They are different. Function composition is not commutative. Problem 2 and 8 provide a counterexample.

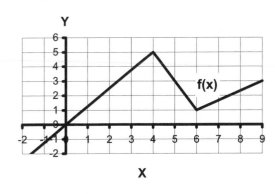
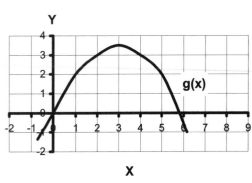

2. Let $f(x) = x^2 + 5$ and $g(x) = \sqrt{x}$.

   (a)   What is the domain of f? **All reals**

(b)   What is the domain of g? $\{x \mid x \geq 0\}$

(c) $f(-1) = $ **6**

(d) $g(f(-1)) = \sqrt{6}$

(e) $f(4) = \mathbf{21}$

(f) $g(f(4)) = \sqrt{21}$

(g) $g(36) = \mathbf{6}$

(h) $f(g(36)) = \mathbf{41}$

(i) $g(7) = \sqrt{7}$

(j) $f(g(7)) = \mathbf{12}$

(k) $f(g(8)) = \mathbf{13}$

(l) $g(f(8)) = \sqrt{69}$

(m)   Why should we not be surprised to find that function composition is not commutative?

Answers will vary. Because the "inside" function contributes to the domain of the composite function, the domains could differ significantly right from the very start, making the function values differ. Since the original function could involve non-commutative operations, applying their definitions in a different order would lead one to suspect the outcomes would be different.

(n) $f(g(-1)) = $ **not defined**

(o) $g(a) = \sqrt{a}$

(p) $f(g(a)) = \mathbf{a + 5}$

(q) $f(p) = \mathbf{p^2 + 5}$

(r) $g(f(p)) = \sqrt{p^2 + 5^2}$

(s) $f(g(x)) = \mathbf{x + 5}$

(t) $g(f(x)) = \sqrt{x^2 + 5}$

(u)   Domain of $f(g(x)) = \{x \mid x \geq 0\}$

(v)   Domain of $g(f(x)) = $ **All reals**

Name _____   Date _____

Function Composition

1.   Refer to the graphs below to approximate each of the following quantities.

    (a)   f(2)

    (b)   g(f(2))

    (c)   f(5)

    (d)   g(f(5))

    (e)   g(3)

    (f)   f(g(3))

    (g)   g(2)

    (h)   f(g(2))

    (i)   Compare the answers from (b) and(h). Is the function composition commutative?
        Explain.

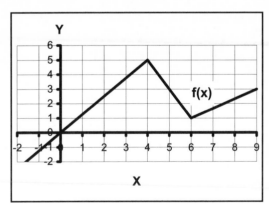
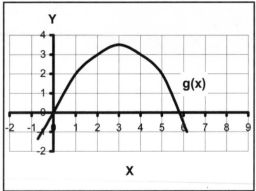

2.   Let $f(x) = x^2 + 5$ and $g(x) = \sqrt{x}$.

    (a)   What is the domain of f?

    (b)   What is the domain of g?

    (c)   f(-1) =

    (d)   g(f(-1)) =

(e)   f(4) =

(f)   g(f(4)) =

(g)   g(36) =

(h)   f(g(36)) =

(i)   g(7) =

(j)   f(g(7)) =

(k)   f(g(8)) =

(l)   g(f(8)) =

(m)   Why should we not be surprised to find that function composition is not commutative?

(n)   f(g(-1))=

(o)   g(a) =

(p)   f(g(a)) =

(q)   f(p) =

(r)   g(f(p)) =

(s)   f(g(x)) =

(t)   g(f(x)) =

(u)   Domain of f(g(x)) =

(v)   Domain of g(f(x)) =

Supplement No. 7-1

Keeping Score in Baseball
(Suitable for use with Section 7.1 of Year 1)

To the Teacher: *This supplement uses baseball to provide practice using set notation.*

1. The positions of baseball players are given in numbers in order to keep score easily. The numbering is as follows:

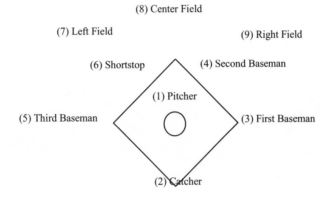

(8) Center Field

(7) Left Field
(9) Right Field

(6) Shortstop
(4) Second Baseman

(1) Pitcher

(5) Third Baseman
(3) First Baseman

(2) Catcher

(a) A = {x | x is an infielder}

   A = {3, 4, 5, 6}

(b) B = {x | x is an outfielder}

   B = {7, 8, 9}

(c) C = {x | x is a member of the battery} Hint: the battery is the pitcher and catcher.

   C = {1, 2}

2. In keeping score, list the elements of the set that are described.

   Ex. A ground ball to the third baseman who throws to first.

   S = {5, 3}

(a) A ground ball to the second baseman who throws to the shortstop covering second base who then throws to first where the first baseman is covering.

   S = {4, 6, 3}

(b) A fly ball to the center fielder who then throws to third base to catch the runner who left the base.

S = {8, 5}

(c)   A bunt to the pitcher who throws to second base to catch the runner and then the second baseman throws to first base to put the runner out.

S = {1, 4, 3}

(d)   A line-drive to the first baseman who throws to the third baseman covering third base who throws to the shortstop covering second base.

S = {3, 5, 6}

(e)   A ground ball to the first baseman who throws to the shortstop covering second base who throws to the pitcher covering first base.

S = {3, 6, 1}

Name _____   Date _____

Supplement No. 7-1

Keeping Score in Baseball

1. The positions of baseball players are given in numbers in order to keep score easily. The numbering is as follows:

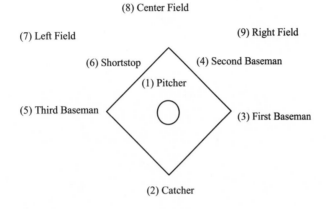

(a)   A = {x | x is an infielder}

(b)   B = {x | x is an outfielder}

(c)   C = {x | x is a member of the battery} Hint: the battery is the pitcher and catcher.

2. In keeping score, list the elements of the set that are described.

   Ex. A ground ball to the third baseman who throws to first.

   S = {5, 3}

(a)   A ground ball to the second baseman who throws to the shortstop covering second base who then throws to first where the first baseman is covering.

(b)   A fly ball to the center fielder who then throws to third base to catch the runner who left the base.

(c)   A bunt to the pitcher who throws to second base to catch the runner and then the second baseman throws to first base to put the runner out.

(d)   A line-drive to the first baseman who throws to the third baseman covering third base who throws to the shortstop covering second base.

(e)   A ground ball to the first baseman who throws to the shortstop covering second base who throws to the pitcher covering first base.

Supplement No. 7-2

Set Notation
(Suitable for use with Section 7.2 of Year 1)

To the Teacher: *This supplement gives students practice identifying the number of elements in different sets.*

1. You are selecting from the numbers 1-40 inclusive (integers only).

    (a) If a is the set of numbers less than 12, what is the #(a)?
    11

    (b) If b is the set of numbers less than or equal to 12, what is the #(b)?
    12

    (c) If c is the set of prime numbers. What is the #(c)?
    12

    (d) If d is the set of numbers between 10 and 30 inclusive, what is the #(d)?
    21

    (e) If e is the set of numbers that are multiples of 2, what is the #(e)?
    20

    (f) If f is the set of numbers that are multiples of 3, what is the #(f)?
    13

    (g) If g is the set of numbers that are multiples of 2 or 3, what is the #(g)?
    27

    (h) If h is the set of numbers that are multiples of 2 and 3, what is the #(h)?
    6

2. A standard deck of card contains 52 cards (no jokers). There are 4 suits (spades, clubs, diamonds, and hearts) and each suit has 13 cards (2, 3, 4, 5, 6, 7, 8, 9, 10, jack, queen, king, and ace). The red cards are hearts and diamonds and the black cards are spades and clubs.

    (a) If a is the number or jacks, what is the #(a)?
    4

    (b) If b is the number of red cards, what is the #(b)?
    26

    (c) If c is the number of black 7's, 8's or 9's, what is the #(c)?
    6

(d) If *d* is the number of red aces or black queens, what is the #(*d*)?
4

(e) If *e* is the number of diamonds or aces, what is the #(*e*)?
16

(f) If *f* is the number of red aces or cards less than 7, (aces are higher than 7) then what is the #(*f*)?
22

3. Use the standard alphabet for the next set of questions.

(a) If *a* is the set of all the vowels, then what is the #(*a*)?
5

(b) If *b* is the set of all non-vowels (consonants), what is the #(*b*)?
21

(c) If *c* is the set of letters without a curve of any type in it, what is the #(*c*)? Use upper-case letters
15

(d) If *d* is the set of letters with a dot in the letter, what is the #(*d*)? Use lower-case letters
2

(e) Repeat question (d) using upper-case letters.
0

Name _____   Date _____

Supplement No. 7-2

Set Notation

1. You are selecting from the numbers 1-40 inclusive (integers only).

    (a) If *a* is the set of numbers less than 12, what is the #(*a*)?

    (b) If *b* is the set of numbers less than or equal to 12, what is the #(*b*)?

    (c) If *c* is the set of prime numbers. What is the #(*c*)?

    (d) If *d* is the set of numbers between 10 and 30 inclusive, what is the #(*d*)?

    (e) If *e* is the set of numbers that are multiples of 2, what is the #(*e*)?

    (f) If *f* is the set of numbers that are multiples of 3, what is the #(*f*)?

    (g) If *g* is the set of numbers that are multiples of 2 or 3, what is the #(*g*)?

    (h) If *h* is the set of numbers that are multiples of 2 and 3, what is the #(*h*)?

2. A standard deck of card contains 52 cards (no jokers). There are 4 suits (spades, clubs, diamonds, and hearts) and each suit has 13 cards (2, 3, 4, 5, 6, 7, 8, 9, 10, jack, queen, king, and ace). The red cards are hearts and diamonds and the black cards are spades and clubs.

    (a) If *a* is the number or jacks, what is the #(*a*)?

    (b) If *b* is the number of red cards, what is the #(*b*)?

    (c) If *c* is the number of black 7's, 8's or 9's, what is the #(*c*)?

    (d) If *d* is the number of red aces or black queens, what is the #(*d*)?

    (e) If *e* is the number of diamonds or aces, what is the #(*e*)?

(f)   If f is the number of red aces or cards less than 7, (aces are higher than 7) then what is the #(f)?

3.   Use the standard alphabet for the next set of questions.

(a)   If a is the set of all the vowels, then what is the #(a)?

(b)   If b is the set of all non-vowels (consonants), what is the #(b)?

(c)   If c is the set of letters without a curve of any type in it, what is the #(c)? Use upper-case letters

(d)   If d is the set of letters with a dot in the letter, what is the #(d)? Use lower-case letters.

(e)   Repeat question (d) using upper-case letters.

Supplement No. 7-3

Venn Diagrams
(Suitable for use with Section 7.3 of Year 1)

To the Teacher: *This Supplement provides practice counting the number of elements in a given set.*

1. Lunch at Bryn's Brunch offers two different meat choices – bacon and sausage. A group of 45 students went into Bryn's and 31 had bacon, 16 had sausage and 6 had both. How many had neither bacon nor sausage? Make a Venn Diagram to support your answer.

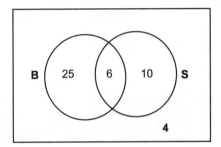

2. A second group of 52 students went into Bryn's with the same two choices of meats to order. 24 students had bacon, 8 had both bacon and sausage and 6 had neither. How many students had sausage only? Make a Venn Diagram to support your answer.

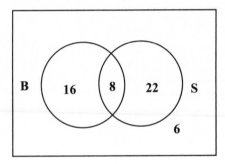

3. Use the Venn diagram to answer the questions below:

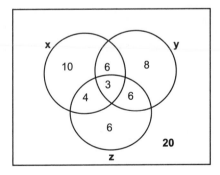

(a)   # y = **23**

(b)   # x = **23**

(c)   # z = **19**

(d)   # (x 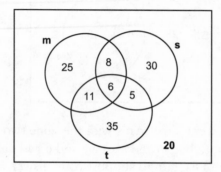 y) = **37**

(e)   # (y ▬ z) = **9**

(f)   # (x 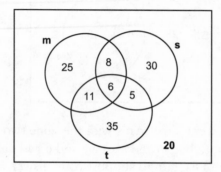 y 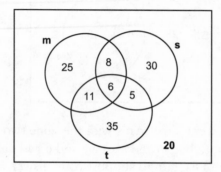 z) = **43**

(g)   # (x ▬ y ▬ z) = **3**

(h)   # x − #(y ▬ z) = **14**

4.   A survey of 140 students was done on what kind of candy bars they liked. Students selected from Snickers, Milky Ways, and Twix. Students could choose more than one candy bar or none of the three.

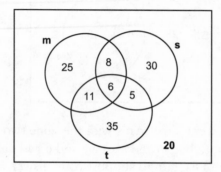

50 liked Milky Ways
49 liked Snickers
57 liked Twix
14 liked Milky Ways and Snickers
17 liked Milky Ways and Twix
11 liked Snickers and Twix
6 liked all 3

Make a Venn diagram to find out how many liked none of the three.

5.   54 people were surveyed on candidates A and B.

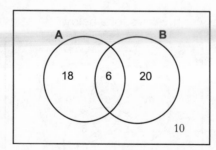

24 liked candidate A
26 liked candidate B
10 liked neither candidate.
How many liked both candidates? Explain how you found your answer.

Name _____   Date _____

Venn Diagrams

1. Lunch at Bryn's Brunch offers two different meat choices – bacon and sausage. A group of 45 students went into Bryn's and 31 had bacon, 16 had sausage and 6 had both. How many had neither bacon nor sausage? Make a Venn Diagram to support your answer.

2. A second group of 52 students went into Bryn's with the same two choices of meat to order. 24 students had bacon, 8 had both bacon and sausage and 6 had neither. How many students had sausage only? Make a Venn Diagram to support your answer.

3. Use the Venn diagram to answer the questions below.

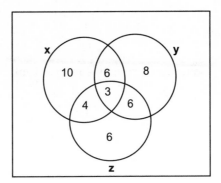

(a) # y =

(b) # x =

(c) # z =

(d) # (x ▮ y) =

(e) # (y ▬ z)

(f) # (x ▮ y ▮ z)

(g)   # (x ■ y ■ z)

(h)   # x − #(y ■ z)

4.   A survey of 140 students was done on what kind of candy bars they liked best. Students
      selected from Snickers, Milky Ways, and Twix. Students could choose more than one candy
      bar or none of the three.

                50 liked Milky Ways
                49 liked Snickers
                57 liked Twix
                14 liked Milky Ways and Snickers
                17 liked Milky Ways and Twix
                11 liked Snickers and Twix
                6 liked all 3

      Make a Venn diagram to find out how many liked none of the three.

5.   54 people were surveyed on candidates A and B.

                24 liked candidate A
                26 liked candidate B
                10 liked neither candidate.

      How many liked both candidates? Explain how you found your answer.

Supplement No. 7-4

Counting Beyond
(Suitable for use with Section 7.5 of Year 1)

To the Teacher: *This Supplement allows the student to practice using the fundamental counting principle.*

1. The state of New Jersey has license plates that contain three numbers followed by three letters. A new policy will allow license plates to have two numbers followed by four letters. How many more license plates can be made by using the two numbers followed by the four letters system than the three numbers followed by the three letters system?

    $26^4 \cdot 10^2 - 26^3 \cdot 10^3 = 28,121,600$

2. How many different combinations can be made if three fair dice are rolled?

    $6^3 = 216$

3. A package of Necco wafers contains forty wafers and there are eight different flavors or colors available to use. How many different arrangements can be made with the forty wafers? (This answer will be huge.)

    $8^{40} = 1.329 \cdot 10^{36}$

4. A package of Skittles has five different fruit flavors in it. If you take five Skittles out of the package, one of each flavor, how many different ways could you arrange these?

    $5! = 120$

5. How many different fruit flavor combinations can you make with the five Skittles? (You may use one Skittle, two Skittles, three Skittles, etc. for a combination. This is a counting problem, not an F.C.P problem.)

    $2^5 - 1 = 31$

6. Please make up a license plate pattern that will make between 250,000 and 350,000 license plates. Example: two letters followed by one digit will make $26 \cdot 26 \cdot 10$ or 6760 plates.

    Answers will vary but one possible solution is 1 letter followed by 4 digits:
    $26 \cdot 10^4 = 260,000$

7.   The famous designer Hilly Tomfigure came out with a new summer fashion of clothes. They have four new types of shirts, three new types of shorts and five new types of socks. How many different combinations of the new summer fashion of clothes can there be?

$4 \cdot 3 \cdot 5 = 60$

8.   The sophomore class at Clishton High School has 435 students. An election for class officers is being held. The positions available are president, vice president. treasurer, and secretary. Each officer can only hold one position. How many different ways could the sophomore class elect the officers?

$435 \cdot 434 \cdot 433 \cdot 432 = 3.531 \cdot 10^{10}$

Name _____ Date _____

Counting Beyond

1.   The state of New Jersey has license plates that contain three numbers followed by three letters. A new policy will allow license plates to have two numbers followed by four letters. How many more license plates can be made by using the two numbers followed by the four letters system than the three numbers followed by the three letters system?

2.   How many different combinations can be made if three fair dice are rolled?

3.   A package of Necco wafers contains forty wafers and there are eight different flavors or colors available to use. How many different arrangements can be made with the forty wafers? (This answer will be huge)

4.   A package of Skittles has five different fruit flavors in it. If you take five Skittles out of the package, one of each flavor, how many different ways could you arrange these?

5.   How many different fruit flavor combinations can you make with the five Skittles? (You may use one Skittle, two Skittles, three Skittles, etc. for a combination. This is a counting problem, not an F.C.P problem.)

6.   Please make up a license plate pattern that will make between 250,000 and 350,000 license plates. Example: two letters followed by one digit will make $26 \cdot 26 \cdot 10$ or 6760 plates.

7.   The famous designer Hilly Tomfigure came out with a new summer fashion of clothes. They have four new types of shirts, three new types of shorts and five new types of socks. How many different combinations of the new summer fashion of clothes can there be?

8.   The sophomore class at Clishton High School has 435 students. An election for class officers is being held. The positions available are president, vice president. treasurer, and secretary. Each officer can only hold one position. How many different ways could the sophomore class elect the officers?

Supplement No. 8-1

Probability
(Suitable for use with Section 8.1 of Year 1)

To the Teacher: *Allow students to express the reasons for their answers. There will be more than one valid response.*

For each of the following events select a number between 0 and 1.0 which indicates the probability you would assign to the given event. Justify your answer.

1.	The New York Yankees will win on opening day.

	P(e) = answers will vary

2.	The stock market will go up.

	P(e) = answers will vary

3.	Mankind will land on Mars.

	P(e) = answers will vary

4.	The President of the United States will be a female.

	P(e) = answers will vary

5.	McDonald's will be the top selling fast food restaurant chain next year.

	P(e) = answers will vary

6.	There will be a snow day tomorrow.

	P(e) = answers will vary

7.	A new red corvette will be bought tomorrow.

	P(e) = answers will vary

8.	What is the probability a baby will be born in a winter month?

	P(e) = $\frac{3}{4}$ = .75

9.	What is the probability that Christmas will fall on a Tuesday?

	P(e) = $\frac{1}{7}$ = .143

10.	What is the probability that a person's first name begins with a J?

	P(e) = 1/26 = .038

11. What is the probability that a phone number ends in the number 2?

    $P(e) = \frac{1}{10} = .10$

12. What is the probability that a person born in the United States was born in New Mexico?

    $P(e) = \frac{1}{50} = .02$

13. What is the probability that a person born in January was born on a prime number day?

    $P(e) = \frac{11}{31} = .355$

Name _____ Date _____

Probability

For each of the following events select a number between 0 and 1.0 which indicates the probability you would assign to the given event. Justify your answer.

1. The New York Yankees will win on opening day.

2. The stock market will go up.

3. Mankind will land on Mars.

4. The President of the United States will be a female.

5. McDonald's will be the top selling fast food restaurant chain next year.

6. There will be a snow day tomorrow.

7. A new red corvette will be bought tomorrow.

8. What is the probability a baby will be born in a winter month?

9. What is the probability that Christmas will fall on a Tuesday?

10. What is the probability that a person's first name begins with a J?

11. What is the probability that a phone number ends in the number 2?

12. What is the probability that a person born in the United States was born in New Mexico?

13. What is the probability that a person born in January was born on a prime number day?

Supplement No. 8-2

Probability
(Suitable for use with Section 8.1 of Year 1)

To the Teacher: *Students will have experience with determining probability in several manageable situations.*

1. There are three red gold balls and seven white golf balls in a golf bag.

    (a) What is the probability of picking a red golf ball

    $$P(e) = \frac{3}{10} = .30$$

    (b) What is the probability of picking a white golf ball?

    $$P(e) = \frac{7}{10} = .70$$

    (c) How many white golf balls must be added so that the probability of picking a white golf ball is .80?

    $$P(e) = \frac{7+5}{10+5} = \frac{12}{15} = .80 \quad \textbf{5 white balls}$$

    (d) How many red golf balls must be added so that the probability of picking a red golf ball is .50?

    $$P(e) = \frac{3+4}{10+4} = \frac{7}{14} = .50 \quad \textbf{4 red balls}$$

2. There are four blue pens, seven black pens, and thirteen red pens on a desk.

    (a) What is the probability of picking a red pen?

    $$P(e) = \frac{13}{24} = .542$$

    (b) What is the probability of picking a blue pen?

    $$P(e) = \frac{4}{24} = \frac{1}{6} = .167$$

    (c) What is the probability of picking a black pen?

    $$P(e) = \frac{7}{24} = .292$$

(d) How many pens of each color need to be added to end up with the following probabilities for each color pen? Hint: Total number of pens < 48.

(1.) Blue Pen = $\frac{1}{6}$

$$P(e) = \frac{4+2}{24+12} = \frac{6}{36} = \frac{1}{6}$$ **2 blue pens and 10 other colors**

(a total of 12 new pens)

(2.) Black Pen = $\frac{1}{3}$

$$P(e) = \frac{7+5}{24+12} = \frac{12}{36} = \frac{1}{3}$$ **5 black pens and 10 other colors**

(a total of 12 new pens)

(3.) Red Pen = $\frac{1}{2}$

$$P(e) = \frac{13+5}{24+12} = \frac{18}{36} = \frac{1}{2}$$ **5 red pens and 10 other colors**

(a total of 12 new pens)

Supplement No. 8-2

Probability

1.   There are three red gold balls and seven white golf balls in a golf bag.

   (a)   What is the probability of picking a red golf ball

   (b)   What is the probability of picking a white golf ball?

   (c)   How many white golf balls must be added so that the probability of picking a white golf ball is .80?

   (d)   How many red golf balls must be added so that the probability of picking a red golf ball is .50?

2.   There are four blue pens, seven black pens, and thirteen red pens on a desk.

   (a)   What is the probability of picking a red pen?

   (b)   What is the probability of picking a blue pen?

   (c)   What is the probability of picking a black pen?

(d) How many pens of each color need to be added to end up with the following probabilities for each color pen? Hint: Total number of pens < 48.

(1.) Blue Pen = $\frac{1}{6}$

(2.) Black Pen = $\frac{1}{3}$

(3.) Red Pen = $\frac{1}{2}$

Supplement No. 8-3

Probability
(Suitable for use with Section 8.2 of Year 1)

To the Teacher: *This supplement provides an opportunity for the student to analyze tossed coins. The formula will be helpful when the number of coins rises.*

Formula: $P(x = k) = \binom{N}{K} p^K q^{N-K}$

1.   A fair coin is tossed twice.

    (a)   List all the possibilities for heads.

        HH, HT, TH, TT

        P(2H) = $\dfrac{1}{4}$ **P(1H) = $\dfrac{1}{2}$**   **P(0H) = $\dfrac{1}{4}$**

    (b)   Calculate the answer to part (a) with the above given formula.

        $\mathbf{P(2H) = \binom{2}{2}(.5)^2 (.5)^0 = \tfrac{1}{4}}$

        $\mathbf{P(1H) = \binom{2}{1}(.5)^1 (.5)^1 = \tfrac{1}{2}}$

        $\mathbf{P(0H) = \binom{2}{0}(.5)^0 (.5)^2 = \tfrac{1}{4}}$

    (c)   Draw a graph of your results

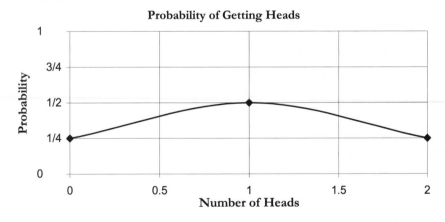

2. A fair coin is tossed three times.

   (a) List all the possibilities for heads.

   HHH, HHT, HTH, HTT
   THH, THT, TTH, TTT

   (b) Calculate the answer to part (a) with the formula given earlier.

   $$\mathbf{P(3H)} = \binom{3}{3}(.5)^3(.5)^0 = \tfrac{1}{8} \qquad \mathbf{P(2H)} = \binom{3}{2}(.5)^2(.5)^1 = \tfrac{3}{8}$$

   $$\mathbf{P(1H)} = \binom{3}{1}(.5)^1(.5)^2 = \tfrac{3}{8} \qquad \mathbf{P(0H)} = \binom{3}{0}(.5)^0(.5)^3 = \tfrac{1}{8}$$

   (c) Draw a graph of your results.

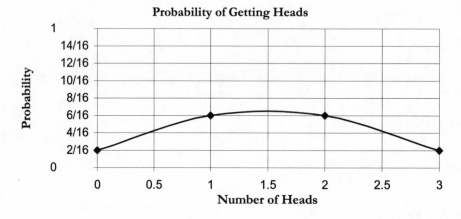

3. A fair coin is tossed four times.

   (a) List all the possibilities for heads.

   HHHH, HHHT, HHTH, HHTT,
   HTHH, HTHT, HTTH, HTTT
   THHH, THHT, THTH, THTT
   TTHH, TTHT, TTTH, TTTT

   (b) Calculate the answer to part (a) with the formula given earlier.

   $$\mathbf{P(4H)} = \binom{4}{4}(.5)^4(.5)^0 = .0625 \text{ or } \frac{1}{16}$$

   $$\mathbf{P(3H)} = \binom{4}{3}(.5)^3(.5)^1 = \frac{4}{16}$$

   $$\mathbf{P(2H)} = \binom{4}{2}(.5)^2(.5)^2 = \frac{6}{16}$$

$$P(1H) = \binom{4}{1}(.5)^3(.5)^1 = \frac{4}{16}$$

$$P(0H) = \binom{4}{0}(.5)^0(.5)^4 = \frac{1}{16}$$

(c)   Draw a graph of your results.

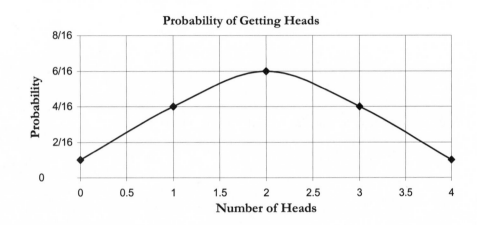

Supplement No. 8-3

Probability

Formula: $P(x = k) = \binom{N}{K} p^K q^{N-K}$

1.   A fair coin is tossed twice.

    (a)   List all the possibilities for heads.

    (b)   Calculate the answer to part (a) with the above given formula.

    (c)   Draw a graph of your results.

2. A fair coin is tossed three times.

    (a) List all the possibilities for heads.

    (b) Calculate the answer to part (a) with the formula given earlier.

    (c) Draw a graph of your results.

3. A fair coin is tossed four times.

    (a) List all the possibilities for heads.

    (b) Calculate the answer to part (a) with the formula given earlier.

(c) Draw a graph of your results.

Supplement No. 8-4

Probability
(Suitable for use with Section 8.2 of Year 1)

Teacher Commentary: *This supplement gives students more experience with probability functions.*

Given the numbers 1 - 30 , inclusive, answer the following questions:

(answers will vary; examples given)

1. Describe an event E in which P(E) = 0.5

    P(odd #) or P(even #) = $\frac{15}{30}$

2. Describe an event E in which P(E) = 0.2

    P(multiple of 5) = $\frac{6}{30}$

3. Describe an event E in which P(E) = 1/3

    P(multiple of 3)   or   P(prime number) = $\frac{10}{30}$

4. Describe an event E in which P(E) = 2/3

    P(not prime number) = $\frac{20}{30}$

5. Describe an event E in which P(E) = 0.3

    P(1-digit #) = $\frac{10}{30}$

6. Describe an event E in which P(E) = 0.7

    P(2-digit #) = $\frac{21}{30}$

7. Describe an event E in which P(E) = .033

    P(E) = .033 = $\frac{1}{30}$
    P(1) = P(8) = P(23) = P(any #)

8. Describe an event E in which P(E) = 0.8

    P(not multiple of 5) = $\frac{24}{30}$ = 0.8

Name _____   Date _____

Probability

Given the numbers 1 - 30 , inclusive, answer the following questions:

1.   Describe an event E in which P(E) = 0.5

2.   Describe an event E in which P(E) = 0.2

3.   Describe an event E in which P(E) = 1/3

4.   Describe an event E in which P(E) = 2/3

5.   Describe an event E in which P(E) = 0.3

6.   Describe an event E in which P(E) = 0.7

7.   Describe an event E in which P(E) = .033

8.   Describe an event E in which P(E) = 0.8

Supplement 8-5

Probabilities: Pies and Games
(Suitable for use with Section 8.3 of Year 1)

Teacher Commentary: *This supplement gives the students more experience working with probabilities.*

1. Given the following multi-flavor pie, answer these questions:

(a) $P(\text{cherry}) = \dfrac{2}{8} = \dfrac{1}{4} = .25$

(b) $P(\text{lime}) = \dfrac{4}{8} = 0.5$

(c) $P(not \text{ cherry}) = \dfrac{6}{8} = .75$

(d) $P(\text{pumpkin}) = \dfrac{0}{8} = 0$

(e) Which is more likely: P(apple) or P(banana)? Explain.

   Same.   P(apple) = P(banana) $= \dfrac{1}{8}$

(f) Which is more likely: P(*not* lime) or P(*not* cherry) ? Explain.

   P(not lime) = $\dfrac{4}{8}$ P(not cherry) = $\dfrac{6}{8}$ P(not cherry) is more likely

2. Create a pie (as in problem #1) which has the following probabilities:

   Pies will vary.

   $P(\text{cherry}) = \dfrac{2}{10}$ **2 cherry slices**

   $P(\text{lime}) = \dfrac{3}{10}$ **3 lime slices**

P(*not* lemon) = $\frac{9}{10}$ **1 lemon slice**

P(apple) = $\frac{4}{10}$ **4 blue slices**

P(*not* mincemeat) = 1   **no mincemeat slices**

3.   Create a pie which has the following flavor probabilities:

P(banana) = $\frac{1}{6}$ $\frac{1}{6} = \frac{2}{12}$

                             2 banana slices

P(apple) = $\frac{1}{3}$ $\frac{1}{3} = \frac{4}{12}$

                             4 apple slices

P(*not* pumpkin) = $\frac{5}{6}$ $\frac{5}{6} = \frac{10}{12}$

                             2 pumpkin slices

P(blueberry) = $\frac{1}{12}$ $\frac{1}{12}$

                             1 blueberry slice

P(*not* lemon) = $\frac{3}{4}$ $\frac{3}{4} = \frac{9}{12}$

                             3 lemon slices

4.   You have a 6-sided die which has the following numbers on it: 2, 3, 6, 8, 10, 12 If you roll this "fair" die exactly once, assign probabilities to each event below:

(a)   P(odd number) = $\frac{1}{6}$

(b)   P(even number) = $\frac{5}{6}$

(c)   P(number > 10) = $\frac{1}{6}$

(d)   P(number < 7) = $\frac{3}{6} = \frac{1}{2}$

(e)   P(*not* 5) = $\frac{6}{6} = 1$

(f) P(prime number) = $\frac{2}{6}$ = $\frac{1}{3}$

5. From a standard deck of 52 cards that have been well-shuffled, one card is drawn face down. Assign probabilities to the following events:

(a) P(queen of spades) = $\frac{1}{52}$

(b) P(queen) = $\frac{4}{52}$ = $\frac{1}{13}$

(c) P(red card) = $\frac{26}{52}$ = $\frac{1}{2}$

(d) P(black face card) = $\frac{6}{52}$ = $\frac{3}{26}$

(e) P(*not* face card) = $\frac{40}{52}$ = $\frac{10}{13}$

Supplement 8-5

Probabilities: Pies and Games

1. Given the following multi-flavor pie, answer these questions:

(a)   P(cherry) =

(b)   P(lime) =

(c)   P(*not* cherry) =

(d)   P(pumpkin) =

(e)   Which is more likely: P(apple) or P(banana)? Explain.

(f)   Which is more likely: P(*not* lime) or P(*not* cherry) ? Explain.

2. Create a pie (as in problem #1) which has the following probabilities:

P(cherry) = $\frac{2}{10}$

P(lime) = $\frac{3}{10}$

P(*not* lemon) = $\frac{9}{10}$

$P(\text{apple}) = \dfrac{4}{10}$

$P(not \text{ mincemeat}) = 1$

3. Create a pie which has the following flavor probabilities:

$P(\text{banana}) = \dfrac{1}{6}$

$P(\text{apple}) = \dfrac{1}{3}$

$P(not \text{ pumpkin}) = \dfrac{5}{6}$

$P(\text{blueberry}) = \dfrac{1}{12}$

$P(not \text{ lemon}) = \dfrac{3}{4}$

4. You have a 6-sided die which has the following numbers on it: 2, 3, 6, 8, 10, 12
    If you roll this "fair" die exactly once, assign probabilities to each event below:

    (a) P(odd number) =

    (b) P(even number) =

    (c) P(number > 10) =

    (d) P(number < 7) =

    (e) P(*not* 5) =

    (f) P(prime number) =

5. From a standard deck of 52 cards that have been well-shuffled, one card is drawn face down.
    Assign probabilities to the following events:

(a)   P(queen of spades) =

(b)   P(queen) =

(c)   P(red card) =

(d)   P(black face card) =

(e)   P(*not* face card) =